MY FIRST COUNTING IN SPORTS BOOK

Jadedra Gilmore-Barber

My First Counting in Sports Book © 2024 Jadedra Gilmore-Barber

All rights reserved. This book or any portion thereof may not be reproduced or used in any manner whatsoever without the express written permission of the publisher except for the use of brief quotations in a book review.

Published by: EarKanDee LLC
For more information: earkandee.educate@gmail.com
ISBN: 979-8-9915315-9-7
www.earkandeeonline.com

"If better is possible, good is not enough."
— Benjamin Franklin

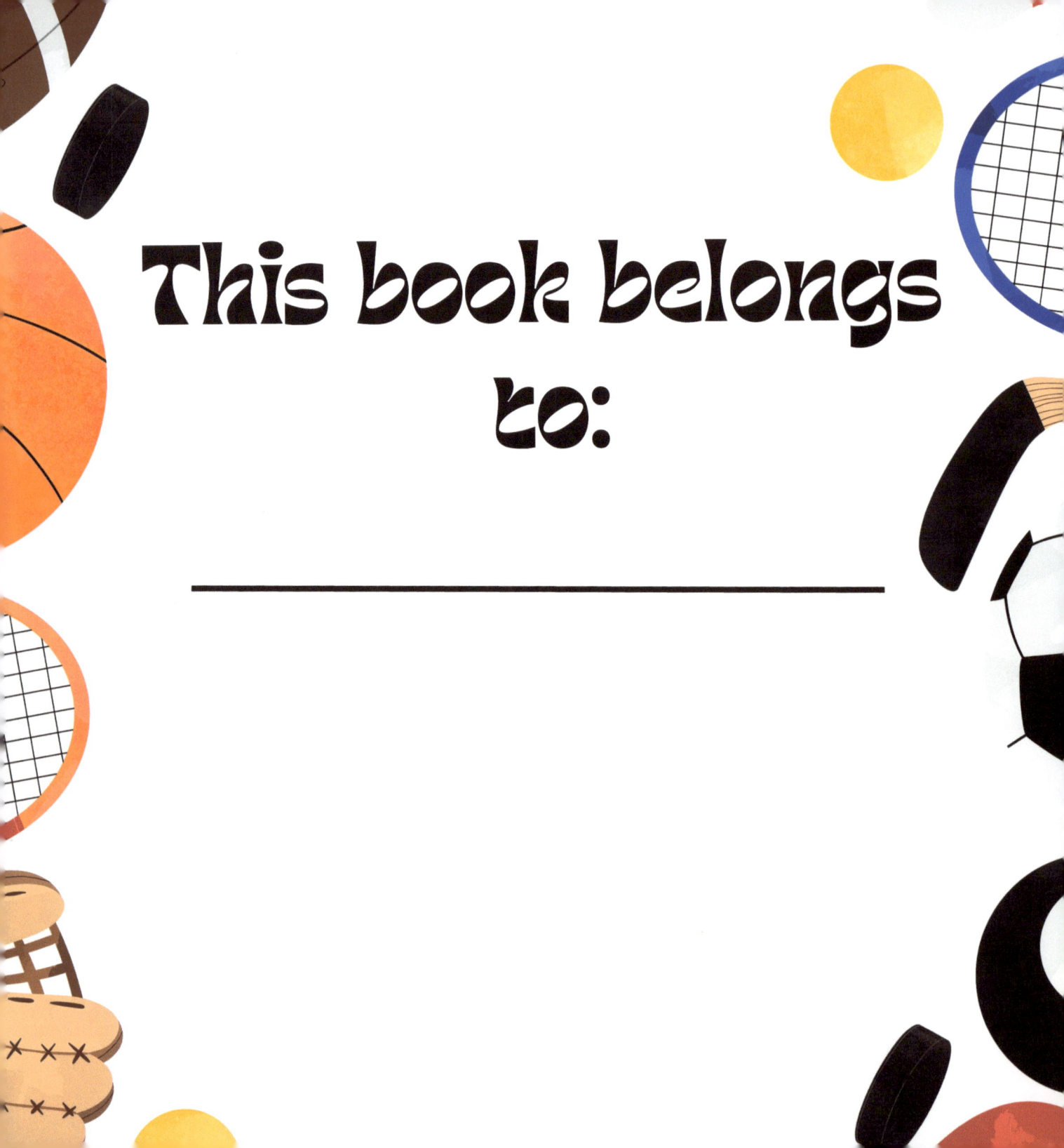

This book belongs to:

_____

TABLE OF CONTENTS

Count Forwards..1
zero..2
one...3
two...4
three...5
four..6
five...7
six..8
seven...9
eight...10
nine..11
ten..12
Count Backwards..13
ten..14
nine..15
eight...16
seven...17
six..18
five...19
four..20
three...21
two...22
one...23
zero..24
WINNER..25

LET'S COUNT FORWARDS!

zero

I see 0 objects.

one

I see _1_ football.

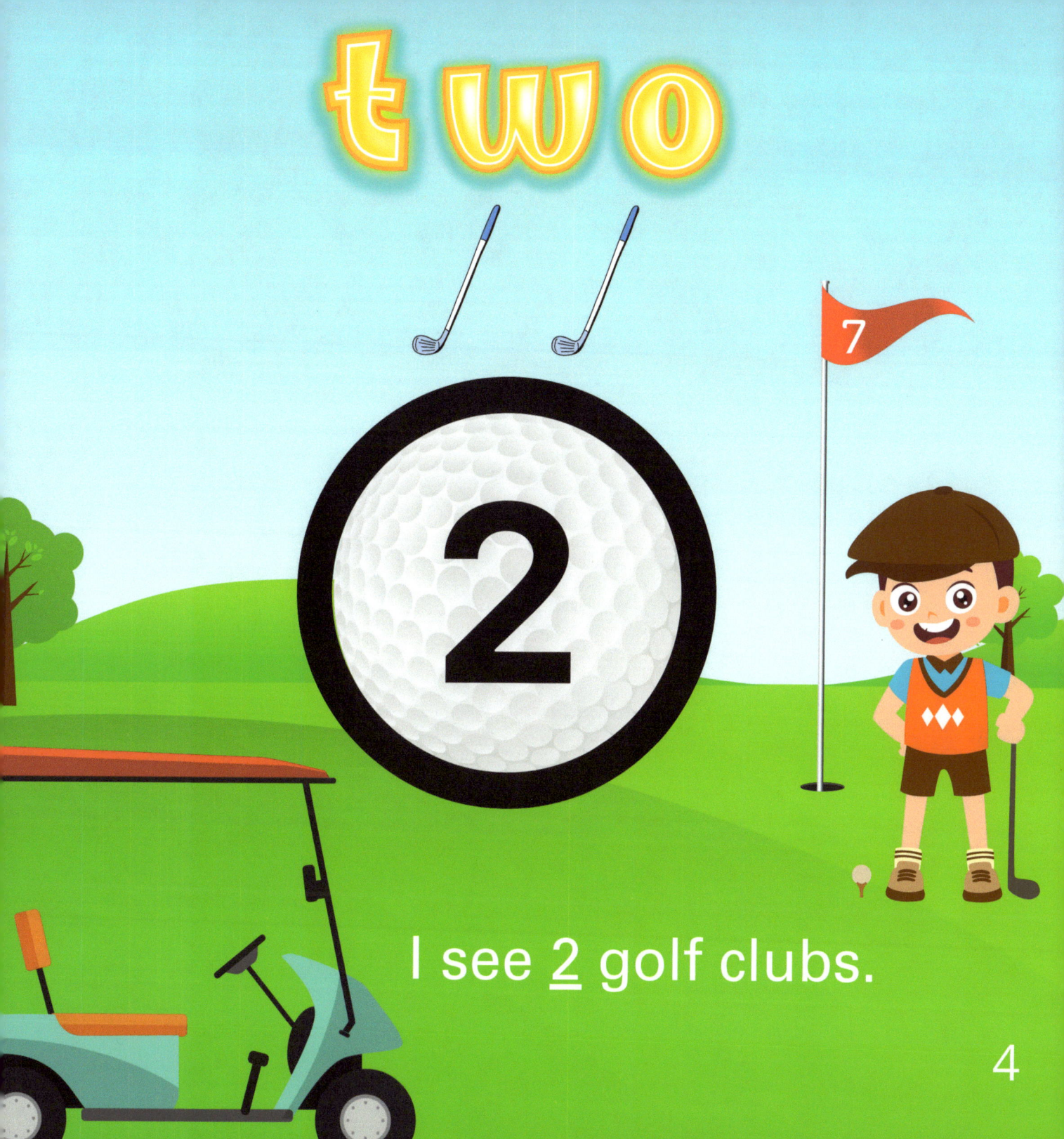

three

3

I see 3 baseballs.

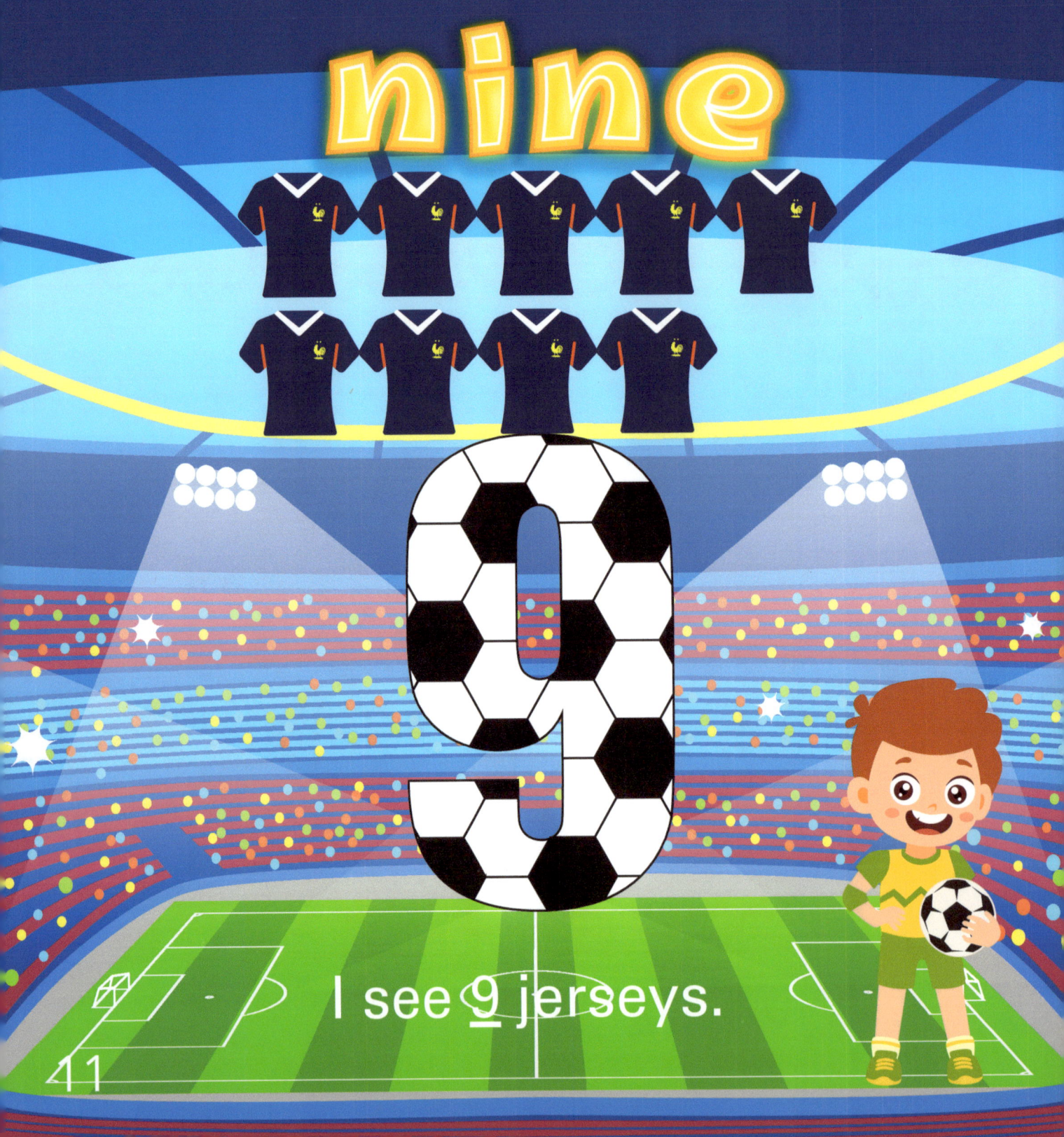

LET'S COUNT BACKWARDS!

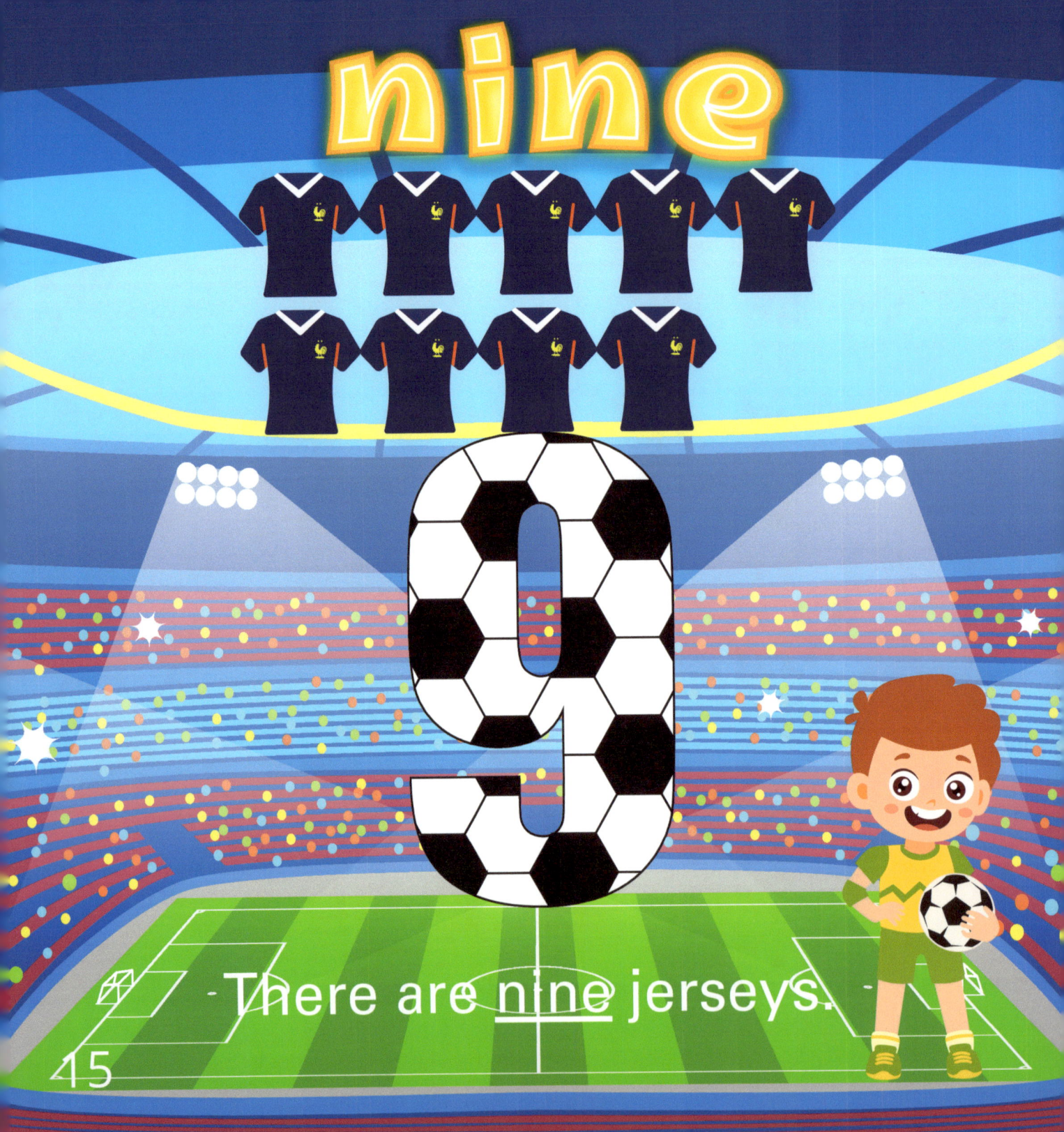

There are four pom-poms.

three

3

There are <u>three</u> baseballs.

one

There is <u>one</u> football.

zero

There are zero objects.

Biography

Dr. Jadedra Gilmore-Barber was born and raised in Fort Valley, Georgia. She has been an educator for over 14 years. Dr. Gilmore-Barber has held various roles during her years in the field of education. She has served as a teacher, academic coach, and administrator. She graduated from Georgia Southwestern State University with a bachelor's degree in Early Childhood Education, a master's degree in Curriculum and Instruction, and a specialist degree in Teacher Leadership. She graduated from Columbus State University with a specialist add-on in Educational Leadership, and a doctoral in Educational Leadership. Dr. Gilmore-Barber's philosophy of education is "Where better is possible, good is not enough!" She strives to motivate students to gain and maintain a love of learning.

EARKANDEE LLC

LISTEN | DESIGN | DISCOVER

"Where better is possible, good is not enough."

PUBLISHING "EFFECTIVE"
EDUCATIONAL RESOURCES GLOBALLY

AVAILABLE NOW ON AMAZON

"MUST LEARN SIGHTWORDS" THROUGH READING, WRITING, AND MATH WORKBOOK (KINDERGARTEN)
AVAILABLE NOW

"MUST LEARN SIGHTWORDS" THROUGH READING, WRITING, AND MATH WORKBOOK (FIRST GRADE)
AVAILABLE NOW

"MUST LEARN SIGHTWORDS" THROUGH READING, WRITING, AND MATH WORKBOOK (SECOND GRADE)
AVAILABLE NOW

MY FIRST COUNTING IN SPACE BOOK
AVAILABLE NOW

MATH "GROWTH" JOURNAL GRADE LEVELS K-5
AVAILABLE NOW

MATH "GROWTH" JOURNAL GRADE LEVELS 6-8
AVAILABLE NOW

MATH "GROWTH" JOURNAL GRADE LEVELS 9-12
AVAILABLE NOW

MY FIRST COUNTING WITH JUNGLE ANIMALS BOOK
AVAILABLE NOW

WRITING ON THE ROADS OF THE USA "STATES": K-2 HANDWRITING WORKBOOK
AVAILABLE NOW

G.R.E.A.T. IS FOR A DAY WITH MY GRANDPARENTS! CHILDREN'S BOOK
AVAILABLE NOW

THE MIGHTY MOVE UPSTAIRS CHILDREN'S BOOK
AVAILABLE NOW

MY FIRST COUNTING WITH MERMAIDS BOOK
AVAILABLE NOW

Stay Tuned For More....

DON'T FORGET TO REVIEW THIS WORKBOOK ON AMAZON

Find us at:
www.earkandeeonline.com

Contact us:
earkandee.educate@gmail.com

www.ingramcontent.com/pod-product-compliance
Lightning Source LLC
Chambersburg PA
CBHW040029050426
42453CB00002B/53